STATE OF ILLINOIS
OTTO KERNER, Governor
DEPARTMENT OF REGISTRATION AND EDUCATION
WILLIAM SYLVESTER WHITE, Director

USES OF LIMESTONE
AND DOLOMITE

J. E. Lamar

DIVISION OF THE
ILLINOIS STATE GEOLOGICAL SURVEY
JOHN C. FRYE, Chief   URBANA

CIRCULAR 321   1961

First printing, 1961
Second printing, with addenda, 1965

Printed by Authority of State of Illinois, Ch. 127, IRS, Par. 58.25.

USES OF LIMESTONE AND DOLOMITE

CONTENTS

USES OF LIMESTONE AND DOLOMITE

J. E. Lamar

ABSTRACT

This report describes briefly the uses of limestone and dolomite and gives chemical and/or physical specifications for each use, where known. More than 70 uses are discussed. A bibliography gives sources of additional data.

INTRODUCTION

This report presents briefly the physical and/or chemical specifications for various uses of limestone and dolomite, information that will aid in the maximum utilization of the limestone and dolomite resources of Illinois. A brief discussion of each use is given, together with general specifications that the stone must meet to be acceptable for the indicated use. A bibliography gives sources of additional data. Mention was found of a number of uses for which no specifications could be located; such uses are simply listed.

This circular is an extensive revision of an earlier report (106) that has been out of print for some years.

The specifications given herein are believed to be reasonably typical, but some may vary, depending on the preference of a user. Also, specifications may be relaxed in certain cases to permit the use of a cheaply available stone from deposits near at hand. In some cases it is difficult to distinguish between uses of limestone and dolomite and those of lime. However, if stone is sold to a consumer for a specific purpose, this is considered a use of limestone or dolomite even though the stone may later be converted into lime in the process of utilization.

An attempt has been made to restrict the use of the term "limestone" to those consolidated rocks that consist principally of the mineral calcite (calcium carbonate). The term "dolomite" is applied, insofar as possible, to consolidated rocks composed principally of the mineral dolomite (calcium magnesium carbonate). However, the data in this report are drawn from diverse writers and, as not all of them have necessarily used the terms as indicated, some laxity in exact usage doubtless results.

Two widely used terms, high-calcium limestone and high-magnesium dolomite, are commonly used industrially without exact definition. High-calcium limestone is generally considered to contain more than 95 percent calcium carbonate, but many, if not most, commercial high-calcium limestones contain more than 97 percent or 98 percent calcium carbonate. High-magnesium dolomite probably generally contains more than 20 percent magnesium oxide or 42 percent magnesium carbonate . The term "high purity dolomite" has been used similarly to describe dolomites containing more than 20 percent magnesium oxide and more than 97 percent magnesium carbonate and calcium carbonate together (152).

[3]

Table 1 shows the quantity and value of limestones and dolomites sold in 1959 according to major uses.

TABLE 1 - LIMESTONE AND DOLOMITE SOLD OR USED BY PRODUCERS
IN THE UNITED STATES IN 1959 (90)

Use	Thousands of short tons	Value thousands of dollars	Average price per ton
Agriculture	20,503	$ 35,665	$ 1.74
Alkali manufacture	3,483	3,954	1.14
Building (dimension) stone			
Rough construction	67	248	3.70
Rough architectural	223	3,150	14.13
Dressed (cut and sawed)	454	16,563	36.48
Rubble	172	518	3.01
Calcium carbide manufacture	834	827†	.99
Cement - portland and natural	84,354	89,947	1.07
Coal mine dusting	526	1,899	3.61
Concrete and road stone	251,787	325,411	1.29
Curbing and flagging	36	214	5.94
Fill material	581	560	.96
Filler (not whiting substitute)			
Asphalt	2,829	6,905	2.44
Fertilizer	464	1,046	2.25
Other	326	1,645	5.05
Filtration	255	445	1.75
Fluxing stone	28,206	40,422	1.43
Glass manufacture	1,317	3,979	3.02
Lime and dead-burned dolomite	19,286	30,034	1.56
Limestone sand	2,293	3,818	1.67
Limestone whiting	698	7,184	10.29
Magnesia (dolomite)*	18	22	1.22
Mineral food	654	3,578	5.47
Mineral (rock) wool	2	4	2.00
Paper manufacture	434	1,190	2.74
Poultry grit	146	1,096	7.50
Railroad ballast	4,589	5,693	1.24
Refractory dolomite	242	441	1.82
Riprap	5,449	6,561	1.20
Sugar refining	856	2,098	2.45
Other and unspecified uses	2,871	6,011	2.09

* Includes stone for refractory magnesia
† Cotter, Perry G., U. S. Bur. Mines, pers. comm., 1961.

ABRASIVE

According to Goudge (65), a moderately finely pulverized soft limestone has been used as an abrasive in a process similar to sandblasting for cleaning soft metal molds, and it is possible it may be used for buffing and cleaning of metal to be electroplated.

ACETIC ACID

The sale of limestone for use in making acetic acid has been reported (35). Lime is used to recover acetic acid as calcium acetate from the destructive distillation of wood (92), and the limestone may have been sold for this purpose. A high-calcium limestone would appear to be desirable.

ACID NEUTRALIZATION

Use. - Iron oxides that form on steel products shaped when the steel is hot are removed by "pickling" the products in a mineral acid, commonly sulfuric acid. The spent liquid from this operation consists essentially of ferrous sulfate and sulfuric acid. In order to dispose of the waste liquor it may be treated with limestone and/or lime to neutralize the acid and precipitate the iron. Limestone also may be used to neutralize other types of acid industrial wastes.

Chemical Specifications. - The rate of reaction of pulverized limestone seems to be approximately inversely proportional to the amount of magnesium carbonate it contains in excess of about 2 percent (74, 75). Dolomite or dolomitic limestone reacts too slowly to be used in the treatment of spent pickle liquors. A limestone containing 95 percent or more calcium carbonate is recommended (48).

Physical Specifications. - Probably a variety of sizes of limestone can be used, depending on the character of the acid to be neutralized and the type of equipment employed. The use of limestone crushed to 1- to 3-mm particles in upflow-type neutralizing systems has been suggested (48); in downflow systems 1- to 3-inch stone is commonly used (48).

Remarks. - The use of ground dolomitic limestone (dolomite) to neutralize acid wastes has been described (84). The wastes resulted from the mining and processing of ores high in iron sulfide and from the production of iron sinter and sulfuric acid from the sulfide. The stone used had 54 percent calcium carbonate and 37 percent magnesium carbonate. All of it passed a 35-mesh sieve and about 40 percent was retained on a 200-mesh sieve. Another grade of stone used was 95 percent through a 200-mesh sieve.

AGGREGATES AND ROAD STONE

Use. - Crushed limestone and dolomite are used as aggregate in portland cement concrete for roads, buildings, and other structures and in combination with bituminous materials for roads and similar construction. They also are used to make base courses for various types of pavements and to make water-bound macadam and soil aggregate roads (see also membrane waterproofing).

Physical Specifications. - Aggregates may be classed as coarse or fine. Coarse aggregate is defined as one predominantly retained on a No. 4 (0.187-inch) sieve, whereas fine aggregate will pass a 3/8-inch sieve, almost entirely pass a No. 4 sieve, and predominantly be retained on a No. 200 sieve (9). The fine aggregate produced by crushing limestone or dolomite quarried from undisturbed consolidated deposits of limestone and dolomite is sometimes referred to as stone screenings (80). Fine aggregate, similarly produced, that has been processed by washing or air separation is known as stone sand (80).

Aggregates for various uses must meet a variety of specifications, among the more important of which are specifications relative to particle size distribution, resistance to wear, soundness, and amount of deleterious substances. The size specifications are not discussed here because they are numerous and varied and because the size of the aggregate is a function of crushing and not an inherent property of stone. The resistance to wear, soundness, and certain deleterious substances are related to the character of the stone, however, and are mentioned briefly below.

Wear. – The resistances of limestone or dolomite to abrasion or wear may be determined by the use of the Los Angeles machine (6). In this test a sample of the stone, weighing between 5,000 and 10,000 grams and graded within certain size limits, is placed in a cylindrical tumbling barrel having a baffle in it, together with a charge of steel balls of specified number and weight. The cylinder is rotated a specific number of times and thereafter the weight of stone retained on a No. 12 sieve is determined. Percentage of wear is calculated by dividing the weight of the material finer than the No. 12 screen by the original weight of the sample.

Soundness. – The soundness test (12) is used to determine the weather resistance of a limestone or dolomite. In some cases actual freezing and thawing tests are made, but the test is probably more commonly made using sodium sulfate or magnesium sulfate solutions. When sodium sulfate is used the test roughly involves soaking samples of stated particle sizes in a saturated solution of sodium sulfate at a temperature of $21°C \pm 1°C$ for between 16 and 18 hours. The samples are then allowed to drain and are dried to constant weight at 105° to 110° C. Each soaking and drying constitute one cycle. The test roughly duplicates freezing and thawing in that, like ice, crystals of sodium sulfate form and grow within the pores of the rock and are responsible for breakage of the stone.

Fine aggregates for the soundness test must pass a 3/8-inch screen and be retained on a No. 50 sieve; particle size distribution between these limits also is specified. A sample of 400 grams or more is used. Coarse aggregate must pass a $2\frac{1}{2}$-inch sieve and be retained on a No. 4 sieve. Various size gradations are permitted. A sample of approximately 6,000 grams is used.

After a specified number of cycles of the soundness test, the sample is washed free from sodium sulfate or magnesium sulfate and dried. Weight loss is determined by the use of suitable sieves and from it a weighted average or "corrected percentage loss" is calculated.

When ledge rock samples are to be tested, the pieces should be of about the same size and shape and weigh roughly 100 grams each. Samples should weigh 5,000 grams ± 2 percent. All fragments which break into three or more pieces are considered to have failed. Percentage weight loss is calculated from the weight of the original sample and the weight of the material which failed.

Other Characteristics. – It is not usually possible to tell with certainty whether a limestone or dolomite will pass the soundness and abrasion tests without actually testing it. However, limestones and dolomites that are light in weight or have a high porosity or water absorption should be regarded with suspicion. An exception is reef-type dolomite which, when unweathered in Illinois, commonly is of superior character with respect to both soundness and wear resistance and yet is quite porous. Impure, especially clayey, dull appearing limestones and dolomites are likely to fail in the tests as are those which break readily when struck with a hammer. Many rocks containing streaks or bands of clay or shale will fail in five cycles of the soundness test. Some chert and some limestones having pronounced banding also fail.

Among the potentially deleterious substances that may occur naturally in some limestone and dolomite deposits, excluding materials from overburden or quarry floor materials, are chert, shale, and clay. The last two substances are most likely to occur as beds or partings between layers of stone or in pockets. According to the Illinois Division of Highways (78), maximum allowable amount of clay lumps is 1.0 percent in fine aggregate for portland cement concrete and 0.25 percent in coarse aggregate. Soft and unsound fragments in the latter, including chert, should not exceed 5 percent. Coarse aggregate for handrail concrete should be "free from chert, flint, limonite, shale, and other kinds of material or substances whose disintegration is accompanied by an increase in volume which may cause spalling of the concrete" (81).

In aggregates for bituminous road surfaces, allowable maxima for clay lumps range from 0.5 to 1.0 percent according to the type of construction, and for shale from 2.0 to 4.0 percent.

Table 2 summarizes abrasion and soundness requirements for aggregates for various purposes as specified by the Illinois Division of Highways. The American Association of State Highway Officials (1) and the American Society for Testing Materials (8) also have set up specifications for road aggregates involving abrasion and soundness.

TABLE 2 - SUMMARY OF ABRASION AND SOUNDNESS REQUIREMENTS FOR
CRUSHED LIMESTONES AND DOLOMITES FOR ILLINOIS ROADS (78)

Type of road construction	Abrasion (*) not more than (percent)	Soundness (†) not more than (percent)
BASE COURSES		
Crushed stone	45	25
Waterbound macadam; coarse aggregate	40	25
Portland cement concrete		
Fine aggregate	--	10
Coarse aggregate	35	15
SURFACE COURSES		
Crushed stone	45	25
Bituminous surface seal coat	40	20
Bituminous surface treatment for crushed stone bases		
Cover and seal coat aggregates	40	20
Bituminous surface dense-graded aggregate type		
Cover and seal coat aggregates	40	20
Bituminous concrete dense-graded aggregate type		
Cover coat aggregate	40	20
Bituminous concrete binder and surface courses, fine dense-graded aggregate type		
Coarse aggregate	35	15
Fine aggregate	--	10

TABLE 2 - Continued

Type of road construction	Abrasion (*) not more than (percent)	Soundness (†) not more than (percent)
SURFACE COURSES		
Bituminous macadam surface course		
Cover coat, keystone, and seal coat aggregate	35	15
Sheet asphalt binder and surface course		
Coarse aggregate	35	15
Fine aggregate	--	10
Portland cement concrete pavement		
Coarse aggregate	35	15
Fine aggregate	--	10

(*) Los Angeles abrasion test. ASTM designation C 131
(†) Sodium sulfate soundness, weighted average loss after 5 cycles. ASTM designation C 88-59T (J. D. Lindsay, personal communication, 1960)

AGRICULTURAL LIMESTONE

Use. - Limestone and dolomite are applied to soils to correct soil acidity, add calcium and magnesium, improve soil structure, and maintain or promote conditions favorable for the utilization of soil nutrients by plants and for the growth of desirable soil organisms. Limestone and dolomite so used are variously referred to as agricultural limestone, agstone, ag lime, or simply lime. Application of agricultural limestone to the soil is called liming.

Chemical Specifications. - The acid neutralizing value of agricultural limestone is of major significance and is measured in terms of the stone's calcium carbonate equivalent, sometimes abbreviated C. C. E., also referred to popularly as its lime content, calcium content, or test. Pure limestone has a C.C.E. of 100 percent, pure dolomite, 108.6 percent.

Physical Specifications. - No generally standard size specifications for agricultural limestone are known. In Illinois, agricultural limestone that will pass an 8-mesh sieve has been recommended (110). Some other states require a more finely ground material.

U. S. Department of Agriculture Specifications. - The Agricultural Stabilization and Conservation Committee of the United States Department of Agriculture has set up the following specifications for agricultural limestone used in connection with its program in Illinois in 1959 (148):

Standard Agricultural Limestone

Ground limestone containing all of the finer particles obtained in the grinding process and ground sufficiently fine so that no less than 80 percent will pass through a United States standard No. 8 sieve.

The calcium carbonate equivalent and the percent passing through a United States standard No. 8 sieve must be at least 80 and one or both must be greater than 80 so that

the multiplication of the percent of calcium carbonate equivalent by the percent of material passing through a United
States standard No. 8 sieve will be equal to or in excess of
0.72. Moisture content at the time of shipment must not exceed 8 percent.

No. 2 Limestone

Ground limestone containing at least 65 percent calcium carbonate equivalent.

Ground sufficiently fine so that 80 percent, including
all of the finer particles obtained in the grinding process, will
pass through a United States standard No. 8 sieve.

The Federal cost-sharing rate for "No. 2 limestone" is less than the cost-
sharing rate for "Standard agricultural limestone."

ALKALI

Use. - Limestone is used in the ammonia soda or Solvay process for the
manufacture of the alkali, soda ash (Na_2CO_3), from either natural or artifically produced salt (NaCl) brines (13). About one (59) to one and one quarter (30) tons of
limestone are used for each ton of sodium carbonate manufactured.

Chemical Specifications. - A high-calcium limestone is required, generally with a silica content of less than 1 percent.

Physical Specifications. - The stone should be between one or two inches
and six inches in diameter (30, 59).

Remarks. - One process of caustic soda (NaOH) manufacture involves treatment of sodium carbonate with milk of lime.

ALUMINUM OXIDE

Use. - Limestone in the form of lime has been used in the extraction of
aluminum oxide from bauxite by the Bayer process (130). The limestone was first
converted to lime which was used to make sodium hydroxide. The extraction of the
alumina is accomplished with this alkali.

This use for limestone is believed to be limited or non-existent in this
country at the present time as sodium hydroxide now is made by the electrolysis
of salt (NaCl) (J. S. Machin, personal communication, 1960).

Chemical Specifications. - A high-calcium limestone containing over 97
percent calcium carbonate and less than 1 percent silica has been used.

ALUMINUM PRODUCTION

In the combination process of extracting aluminum from its ores, limestone
is sintered with the leach residues to form insoluble dicalcium silicate with the
silica present. Alumina and soda in the residues may then be leached out and returned to the production process (18). A high-calcium limestone low in silica
would seem to be a requisite for this purpose.

AMMONIA

Sales of limestone to ammonia works are reported, but the use of the limestone is not specified. It may be burned to lime and used for the recovery of
ammonia from weak ammonia liquors (Frank Reed, personal communication, 1937).

ASPHALT FILLER

Use. - Limestone dust is used as filler for asphalt.

Specifications. - No chemical specifications were noted; presumably the term limestone in specifications includes both the calcitic and dolomitic types. Some asphalt filler probably might be classed as whiting but the color of asphalt filler is not as critical as in the case of most whiting nor is the asphalt filler necessarily as finely pulverized as whiting.

Goldbeck (61) states that asphalt fillers should have a fineness of 80 percent minus 200-mesh. Bowles (27) also sets the fineness at approximately 80 percent minus 200-mesh. The Illinois Division of Highways (79) specifies that asphalt filler be of such fineness that 100 percent will pass a No. 30 sieve, 85 percent a No. 100 sieve, and 65 to 100 percent a No. 200 sieve. Tentative ASTM specifications for mineral filler for sheet asphalt and bituminous concrete pavements (11) call for 100 percent passing a No. 30 sieve, 95 to 100 percent a No. 50 sieve, 90 to 100 percent a No. 100 sieve, and 70 to 100 percent a No. 200 sieve. American Association of State Highway Officials (1) specifications for mineral filler for sheet asphalt and bituminous concrete pavement require 100 percent passing a No. 30 sieve, a total of not less than 95 percent passing a No. 80 sieve, and a total of not less than 65 percent passing a No. 200 sieve.

ATHLETIC FIELD MARKING

The sale of limestone for marking athletic fields has been reported (131). A light-colored stone undoubtedly would be desirable.

BARNSTONE

Use. - Crushed limestone or dolomite referred to as barnstone, barn lime, or barn snow is sprinkled on the floors and walls of stock barns, especially dairy barns, where it serves as a neutralizing agent and absorbent of organic wastes and also gives a clean appearance. Presumably it may also be used as a litter for other purposes.

General Specifications. - No specifications have been noted for barnstone, but stone largely passing 8-mesh or stone considerably more finely ground are both reported to have been sold for this purpose. A white or light-colored stone of at least reasonably high purity is probably desirable.

BRICK GLAZING

The sale of limestone for brick glazing has been reported (33). The limestone probably is ground.

BRICK MAKING

Use. - A "soft plastic limestone of over 90% $CaCO_3$" with about 10 percent of red clay added to it has been used in Cuba for making common brick. Bricks and blocks also have been manufactured from limestone screenings with the addition of weathered lava (150). Mixtures of crushed limestone and marl have been used experimentally for making extruded and dry pressed building brick (114).

Chemical Specifications. - An impure clayey limestone or a mixture of limestone and a clay or clayey material appears desirable.

BULB GROWING

The sale of limestone for bulb growing is reported (47). The limestone probably should have an attractive color and be in small chips.

CALCIUM CARBIDE

Use. - Limestone, converted to lime, is used in the manufacture of calcium carbide. The carbide is made by heating the lime with coke in an electric furnace.

Chemical Specifications. - Chemical specifications (63, 64, 113, 135; R. B. Ladoo, personal communication, 1959) vary but agree that the limestone should contain 97 percent or more calcium carbonate ($CaCO_3$). The maximum phosphorous (P) allowed is 0.01 percent; some specifications have lower maxima of 0.008 and 0.004 percent. Magnesia (MgO) should be less than 2 percent and some specifications called for less than 1 percent. Silica (SiO_2) maxima range between 1 and 3 percent and the maxima for iron oxide (Fe_2O_3) and alumina (Al_2O_3) together vary between 0.05 and 0.75 percent. Sulfur (S) is allowable in trace amounts only.

Physical Specifications. - The limestone must retain its form during burning and yield a tough, strong, lump or pebble lime with a minimum of fines (R. B. Ladoo, personal communication, 1959). Some users specify oolitic limestone.

CALCIUM NITRATE

Calcium nitrate (93, 97) is made by treating limestone with nitric acid. A high-calcium limestone is required.

CARBON DIOXIDE

Use. - Carbon dioxide is obtained from many sources including lime kiln gases (109). The sale of limestone for making carbon dioxide also is reported (89). The stone so used probably is dolomite, and the carbon dioxide is obtained from it by reaction with sulfuric acid (94, 128).

Specifications. - No specifications have been found for stone to be used for the generation of carbon dioxide, but doubtless a high-purity limestone or dolomite would be desirable.

CHROMATE

Dolomite is used in making chromate (129).

COPPER PURIFICATION

The sale, or use by the producer, of limestone for the purification of copper is reported (143). It is probably used as a flux or as a lime in the refining process, particularly as a "slag-forming material in the electrothermic refining of copper" (M. F. Goudge, personal communication, 1937).

DIMENSION STONE

Use. - Limestone and dolomite are used as structural materials and for decoration and monuments. Various terms are applied commercially to stone for specific purposes or to stone having different size and physical characteristics, but the names are not always exactly defined or used. The following general categories are discussed here: building stone, including cut stone, ashlar, rough building stone, rubble and veneering stone, flagging, curbing, and monumental stone. All the grades of stone mentioned fall in the broad category of dimension stone because size or shape limitations are inherent in different degrees in each grade.

Weather Resistance and Tests

In all uses of stone for exterior construction, weather resistance is a requisite property. The ability of limestones and dolomites to withstand the destructive effects of weather depends on many factors, some of which relate to inherent properties of the stone and others, particularly in the case of limestones or dolomites showing textural banding, to whether the stone is used with its bedding in a horizontal or vertical position and to what extent it is protected from the infiltration of water. A few of the more important characteristics of limestones and dolomites that affect their weather resistance are mentioned below.

Clay seams or partings or clay or shale inclusions in limestone or dolomite generally should be regarded as possible causes of splitting when the stone is exposed to the weather. Streaks or bands of pyrite or marcasite may produce the same effect. Chert layers or nodules may cause splitting or spalling but not all chert is equally harmful. Stylolites, also called "crows feet," in some limestones and dolomites may afford access to water and hence facilitate the destructive activities of freezing water. The phenomenon is favored where stone is used in such a position that it is "on edge" with respect to its natural bedding.

An understanding of the distribution and character of impurities in limestones and dolomites is facilitated by etching a flat or smoothed surface with dilute hydrochloric acid. The acid dissolves the carbonate of the rocks but leaves the impurities untouched. Etching with dilute acetic acid will show the probable texture of a surface of the stone after it has weathered for a considerable time (102).

Some limestones and dolomites contain appreciable soluble salts, especially sulfates, which probably occur between the crystalline particles composing the stone (105). If such stones are subjected to wetting and drying, the soluble salts may be brought to the surface and precipitated as a white or whitish surficial discoloration called efflorescence. Weathered natural deposits of stone sometimes show such an efflorescence. An idea of whether a stone will effloresce may be obtained by immersing a dry, brick-size piece of the stone vertically in a dish of distilled water so that the stone is about one half immersed. Usually in the course of a few weeks efflorescence will appear on the specimen above the water, if it is present. It may be necessary to dry the stone after the test to recognize the presence of efflorescence.

Much valuable data regarding the weather resistance and performance of building stone for exteriors may be obtained from the study of the effects of the weather on outcrops of the stone and on actual installations of the stone in old buildings, walls, foundations, chimneys, and the like. It is likely that most limestones and dolomites that have satisfactorily withstood the weather in actual structures, even small structures, will prove satisfactory, if properly used, for other exterior construction. They are, in general, also likely to have ample strength to meet most or all modern requirements for building stone.

Many tests have been proposed to determine the suitability of stone for various uses for construction and decoration, including strength tests, abrasion tests, absorption tests, chemical analyses, microscopic examination, freezing and thawing tests, thermal expansion tests, the combined effect of temperature changes and weak salt solutions, and others (5, 50, 142, 149), but no generally accepted specific limits that distinguish between acceptable and unacceptable building stone are known to have been set up for the tests. However, certain maximum or minimum values may be specified by engineers or architects in connection with individual projects.

Cut Stone

Use. - Limestone and dolomite cut stone is used in the exterior of buildings for walls, sills, trim, and decoration, and in the interior for walls, wainscoting, flooring, and other similar purposes. Some of it is used as monuments or carved for decorative purposes.

Stone for Exterior Use

Physical Specifications. - Cut stone for exterior use should have good weather resistance and a pleasing appearance. It should not contain mineral grains which, when they weather, produce colored streaks and stains unless such discoloration is not considered undesirable. In particular, the minerals pyrite and marcasite (which produce the yellow and brown stains of iron rust) may be undesirable in cut stone for many uses.

Deposits intended to serve as sources of cut stone for exterior use should generally be composed of relatively thick strata free from numerous joints or fractures. It is often desirable that the stone be "free working," that is, that it split with approximately equal ease in all directions. Stone for exterior use usually should not be quarried from or immediately adjacent to sites where stone has been quarried by heavy blasting because the blasting may have produced incipient cracks in the deposit that do not manifest themselves immediately but may appear and cause deterioration when the stone is exposed to the weather.

Stone for Interior Use

Physical Specifications. - Limestone or dolomite for interior use may be polished or used with a smoothed, honed, or tooled surface. The polished stone so employed is known commercially as marble. An attractive appearance and freedom from defects is desirable in all cases. Deposits from which limestone or dolomite for interior use are to be obtained should generally possess the same characteristics as those mentioned for exterior stone, except that weather resistance usually is not as critical.

Stone used for flooring, steps, and the like should be resistant to abrasion. Federal specifications (149) require an abrasion hardness of 12 or more (7). Stone that is subject to frequent washing or wetting should not contain pyrite, marcasite, or clay partings as these materials may cause failure.

Ashlar, Rubble, Veneering, Flagging, and Curbing

Use. - Sawed or hewn stone in comparatively small, squared blocks used for masonry is called ashlar. Rubble is rough, unsquared, or roughly squared stone used for masonry. Comparatively thin slabs of stone used as a facing stone are known as veneering. Flat stones or slabs used for paving walks, patios, and the like are called flags, flagstones, or flagging, and similar flat stones or slabs used for making curbs are known as curbing. Rough building stone consists of pieces of various sizes and shapes.

Physical Specifications. – Ashlar for interior or exterior use should have good weather resistance and essentially the same physical characteristics as cut stone for these purposes.

Rubble and veneering stone should have good weather resistance and at least one face with an attractive appearance. Stone containing substances such as pyrite or marcasite, which produce stains when they weather, should be avoided unless such stains are not undesirable. A relatively thin–bedded deposit is probably desirable for rubble production. For veneering stone, a deposit composed largely of layers roughly 1 to 4 inches thick is commonly desirable. The surfaces of the bedding planes in a veneering stone deposit should have pleasing appearance and color.

Deposits used as a source of flagstone should consist principally of layers a few inches to about six inches thick, and the bedding plane surfaces should be relatively smooth and even. A high degree of weather resistance is a necessity, as is good resistance to abrasion. Deposits from which curbstones are produced also should have a high degree of weather resistance.

Monumental Stone

Use. – Limestone or dolomite is used for markers, headstones, monuments, and the like.

Physical Specifications. – A uniform stone of superior weather resistance and pleasing appearance is desirable.

DISINFECTANTS AND ANIMAL SANITATION

The sale of limestone for disinfectants and animal sanitation has been reported (89). This use may in part be the same as that described under the heading of "Barnstone."

DYES

Use. – Calcium carbonate is used in the process of halogenation employed in the manufacture of dye intermediates (127).

General Chemical Specifications. – Probably a high–calcium limestone is used for the halogenation process.

General Physical Specifications. – Limestone for use as a "base for dyes" should be of such size that it will "all pass through a 20-mesh screen and 97 percent through 100-mesh" (117).

ELECTRICAL PRODUCTS

The sale of limestone for electrical products has been reported (89). No details have been found regarding this use. It may be a use for limestone whiting.

EPSOM SALTS (73)

Production of epsom salts and carbon dioxide is reported from dolomite by a series of processes involving as a primary reaction the treatment of the dolomite with sulfuric acid.

General Chemical Specifications. – A dolomite whose average calcium and magnesium carbonate content totals 99 percent is said to have been used.

General Physical Specifications. – The dolomite reportedly is shipped from the quarry in "one man" lumps but is reduced to 60-mesh at the plant.

EXPLOSIVES (140)

Use. - Limestone is used as a neutralizing agent for acids in some types of blasting explosives.

General Chemical Specifications. - The materials used range from "a rather pure calcium carbonate to marble dust containing as much magnesium as calcium. The impurities in the limestone are of no practical importance in the manufacture of explosives."

Remarks. - "The average dynamite contains less than 1.5 per cent of calcium carbonate but special powders have contained as high as 15 per cent."

FERTILIZER FILLER

Use. - Limestone and dolomite are used as fillers for fertilizers to add weight, reduce caking (25), improve the physical condition of the mixture (83), and to adjust the mixture to the desired ratios of fertilizing elements (49). The limestone and dolomite reduce or eliminate the physiological acidity of the fertilizers and are basic constituents employed to give a "physiologically" basic product (15, 83, 112). Dolomite and limestone are both satisfactory for filler (34).

Chemical Specifications. - A reasonably pure limestone or dolomite is required (E. E. DeTurk, personal communication, 1936).

Physical Specifications. - Usually in the No. 8 to No. 20 sieve sizes (60).

FILL MATERIAL

Limestone or dolomite is used to make fills to serve as a base for various structures. Sometimes stone removed during stripping operations is used for this purpose.

FILTER STONE

Use. - Crushed limestone or dolomite is used in sewage disposal plants to form the beds of trickling filters over which the liquid portion of the sewage is sprayed. The rock serves as a host for organisms which purify the sewage.

Physical Specifications. - Filter stone is used in two sizes, $3\frac{1}{2}$ by $2\frac{1}{2}$ inches and 3 by $1\frac{1}{2}$ inches (61). Careful grading is required,together with close limitations on the amount of fines (61). Siliceous impurities are not objectable if they are fine grained and evenly distributed, but pyrite, marcasite, and clay should be avoided (27). Some types of chert are undesirable, particularly if used in the upper part of the filter bed (101). Filter stone should have a rough surface to provide anchorage for bacteria and other organisms. It generally should withstand 20 cycles of the sodium sulfate soundness test with little or no loss of material (4).

Remarks. - Limestone and dolomite are competitive with granite, quartzite, trap rock, slag, and other materials as filter stone.

FLUX

Limestone and dolomite are used as fluxes in the smelting of metalliferous ores. They combine with impurities in the ores, such as silica and alumina, to form a fluid slag that can be separated from the metal.

Blast Furnace Flux

Use. – Either limestone or dolomite is used as a flux in the production of pig iron from iron ore in blast furnaces. Locally magnesian limestone is used (39).

Chemical Specifications. – Specifications vary, probably because cost and availability of stone may influence the kind of stone used, favoring the use of more impure stone in certain places (34). One specification states that silica should not exceed 2 percent and that sulfur and phosphorous should not be greater than 0.1 percent each (62). A maximum of 0.5 percent is another figure given for sulfur content (28). Other specifications (39) call for less than 5 percent silica, with some plants specifying less than 3 percent, alumina less than 2 percent, magnesia less than 4 to less than 15 percent at various plants, and phosphorous pentoxide not more than a trace (0.005 to 0.006 percent). Still other data (60) give a general maximum of 1.5 percent for silica or a preference for less than 1.0 percent (121).

Physical Specifications. – Size specifications for fluxing stone vary and include ranges of from 1 to 4 inches (60), 2 to 4 inches (41), and 2 inches by one-half inch to 6 by 3 inches for high-calcium limestone and 2 inches by one-half inch to 6 by $1\frac{1}{2}$ inches for dolomite (56). The stone should be strong, and some users specify that it not decrepitate under the heat of the furnace (39).

Remarks. – Blast furnace slag that is to be used for making cement should be low in magnesia, and hence a limestone flux should be used. A figure of less than 3 percent magnesia in such slag is given by Miller (116). For concrete aggregate or railroad ballast a high-magnesia slag is harder and less subject to slaking (41). Seven to 10 percent magnesia is said to be desirable in slag for road making (20).

Open-Hearth Flux

Use. – Limestone is used as a flux in the basic open-hearth process of making steel.

Chemical Specifications. – Limestone containing less than 1.5 percent alumina, 1 percent silica, and 10 percent magnesium carbonate (5 percent magnesia) is used (20). Another figure for maximum silica is 3 percent (63). Still another specification calls for high-calcium limestone low in silica, sulfur, and phosphorous (60). More specific is a figure of 98 percent calcium carbonate as preferable, with phosphorous not to exceed trace amounts (39).

Physical Specifications. – Limestone in 4- to 8-inch pieces is preferred (41); other specifications call for 5 by 8 inches to $5\frac{1}{2}$ to 11 inches (56).

Remarks. – High-calcium limestone flux is used rather than dolomite because calcium oxide has a greater affinity for phosphorous than does magnesia. Phosphorous is the most important material the flux is required to remove (20).

Other Fluxing Uses

High-calcium limestone, and less commonly dolomite, is used as flux-stone in foundry cupolas (60). Specifications generally are the same as for blast furnace flux. A common size for the stone usually is in $1\frac{1}{2}$- to 2-inch pieces (60) or in $1\frac{1}{2}$- to three-quarter inch pieces (98). Limestone, probably high-calcium, also is used in the smelting of lead and copper ores which are acidic (60) and in the smelting of other nonferrous-metal ores (22, 64). Still another fluxing use for limestone or dolomitic limestone is in the direct reduction of iron ore in the rotary kiln (139). Small quantities, "depending upon requirements, are fed into the kiln to arrest the sulfur contained in the ore, and in the solid carbonaceous fuel."

GLASS

Use. - Limestone or dolomite in the raw state, or burned to lime, is an important constituent of the "batch" from which glass is made. Some glass batches contain 20 to 30 percent of limestone or dolomite (44).

Chemical Specifications. - The limestone or dolomite should have uniform composition and high purity. The calcium carbonate content of the limestone probably should exceed 98 percent and the calcium and magnesium carbonate of dolomite 98 percent. Iron oxide should not be more than 0.05 percent and preferably less than 0.02 percent (39). Another specification permits a maximum of 0.3 percent iron for most glass and of 0.03 percent for flint glass (60). A low sulfur and phosphorous content is a requisite, and carbon should be kept to a minimum. Not necessarily applicable in this country but informative are the 1959 British standard specifications for limestone for making colorless glasses (37) which require in part that calcium oxide should not be less than 55.2 percent (98.5 percent calcium carbonate) and total iron as ferric oxide not more than 0.035 percent.

Physical Specifications. - Limestone and dolomite for glass should be crushed to pass a 16- or 20-mesh sieve and should be largely coarser than 100- or 140-mesh. The British specifications mentioned above require that limestone for use in tank furnaces pass a three-sixteenth-inch sieve with not more than 25 percent passing a No. 120 sieve. For pot furnaces the limestone must pass a one-eighth-inch sieve with not more than 5 percent retained on a No. 14 sieve.

LIME

Use. - Lime is made from limestone or dolomite by burning them so as to drive off carbon dioxide. Limestone yields a product consisting mainly of calcium oxide whereas the product from dolomite is mainly calcium and magnesium oxides.

The great bulk of the commercial lime used in the United States consists of calcium oxide and between 0 and 45 percent magnesium oxide and less than 5 percent of silica, alumina, iron oxide, and other impurities. Bowles (26) describes three kinds of lime as follows: high-calcium lime, containing not less than 90 percent calcium oxide and 0 to 5 percent magnesium oxide; dolomitic or high-magnesium lime, containing 25 to 45 percent magnesium oxide; and low-magnesium lime, containing 5 to 25 percent magnesium oxide.

Of the above, the first two are by far the most widely used in the United States; low-magnesium lime is produced in only small tonnages (Robert S. Boynton, personal communication, 1960).

Some limes have the property of setting under water and thus are called hydraulic limes. They are made by burning limestone containing sufficient argillaceous (26) or siliceous (Robert S. Boynton, personal communication, 1960) matter to form a substance that will set under water. Production in the United States is limited but hydraulic lime is produced in quantity in Europe (Robert S. Boynton, personal communication, 1960).

Chemical Specifications. - It is likely that most high-calcium limes are made from limestone containing less than 3 percent impurities, less than 5 percent magnesium carbonate, more than 90 percent calcium carbonate, and often more than 97 or 98 percent calcium carbonate. Similarly, high-magnesium limes (excluding those made from magnesitic dolomite) probably are made from dolomites containing less than 3 percent impurities and more than about 40 percent magnesium carbonate. Low-magnesium limes would presumably be made from dolomite or dolomitic limestone containing less than 3 percent impurities and roughly between 6 and 40 percent magnesium carbonate.

No recent specifications have been found giving data on the maximum amount of argillaceous matter allowable in limestones to be used in making hydraulic lime, but earlier data suggest a maximum of about 25 percent noncarbonate components with magnesium carbonate usually less than 10 percent (106).

Physical Specifications. - Limestone may be burned to lime either in vertical or rotary kilns. For the former between 3- (Robert S. Boynton, personal communication, 1960) and 10-inch (29) stone may be used, but most kilns use 5- to 8-inch stone (36). In rotary kilns three-eighths to $2\frac{1}{2}$-inch stone may be used, although stone of narrower size limits gives better results (36). Minus 6-mesh limestone is used in the fluosolids process of lime burning (26).

Limestone and dolomite should be sufficiently hard so that there is little production of fines or dust during burning and should generally yield a strong firm lime. Some limestones and dolomites decrepitate when heated. This is an undesirable property if pronounced because it increases the amount of fines.

Lime is made from limestones of many textural kinds, ranging from finely to coarsely crystalline. Oolitic limestone is preferred by some producers.

Remarks. - Vienna lime is a kind of lime used for polishing and buffing metal and for giving an "under surface" blue to nickel after plating (52) and to highly polished nickel articles (111). A fine-grained dolomite of high purity that contains 21 percent magnesia and 30 percent lime (CaO) (134) is said to be suitable for making Vienna lime. The analysis of a dolomite used for making Vienna lime (52) shows roughly 55 percent calcium carbonate (31 percent calcium oxide) and 43 percent magnesium carbonate (21 percent magnesium oxide) with traces of silica, iron oxide, and alumina. The grain size, porosity, and crystalline character of the dolomite also may be of importance.

LITHOGRAPHIC STONE

Very pure, uniform, fine-grained limestone or dolomitic limestone, known as lithographic stone, having an exceptionally even texture and free from grit or other granular impurities, was at one time extensively used for making lithographs. The present use of the stone for making lithographs in the art field has been described (76). It also is said to be used in the manufacture of decalcomania (100).

Deposits at Bavaria, Germany, are known widely as sources of lithographic limestone. The stone is used as slabs about 3 inches thick; common sizes range from 16 by 22 inches to 43 by 64 inches (100).

MAGNESIUM AND MAGNESIUM COMPOUNDS

Use. - Dolomite is used in a number of processes for the production of magnesium metal or magnesium compounds (68): (a) it serves as a source of magnesium metal produced by the silicothermic process (45, 46); (b) it is calcined and reacted with sea water or brines to make refractory magnesia and other magnesium compounds (17, 42, 45, 67); (c) it is calcined and washed with carbon dioxide-free water to remove most of the calcium oxide, thus leaving magnesia which is used for refractories; and (e) calcined dolomite is converted to milk of dolomite which is processed by various procedures, including the Pattinson process (118), to yield various magnesium compounds, one of which — basic magnesium carbonate — is employed in the manufacture of "85% magnesia," an insulating product for boilers and steam pipes.

Specifications. - No comprehensive specifications have been noted for dolomite to be used in making magnesium and magnesium compounds, but it is likely that a dolomite of high purity is generally desirable. The dolomite used to treat magnesium bitterns at a California plant making perclase pebbles and light-burned magnesia must contain less than 0.85 percent silica (SiO_2) and 0.16 percent R_2O_3 (G. W. Martin, personal communication, 1960). A dolomite used in Pennsylvania for making magnesium carbonate for insulation contained 40 to 45 percent magnesium carbonate and less than 1 percent silica (118) and was in 8- to 12-inch pieces. The size of stone used in the various processes mentioned undoubtedly varies.

MAGNESIUM CHLORIDE

A process for making magnesium chloride from Texas dolomite was developed in 1945 (156). The dolomite used contained approximately 42 percent $MgCO_3$, 55 percent $CaCO_3$, and less than 3 percent combined SiO_2 and R_2O_3.

MASONRY CEMENT

Use. - Masonry cements (122) are prepared by "intergrinding portland cement clinker or finished portland cement with limestone and an air entraining plasticizer to a fineness greater than that of portland cement." Another variety of masonry cement is made by grinding together natural cement and a small amount of portland cement, and sometimes, in addition, a small amount of limestone. Masonry cement also is prepared from slag-lime cement and a small quantity of portland cement.

No specifications have been found for the limestone ground with the other components of masonry cements.

MEMBRANE WATERPROOFING

Use. - Membrane waterproofing, placed on various highway structures, consists of three layers of bitumen-treated cotton fabric (mopped with bitumen and given a protective cover). One kind of cover consists of a mixture of petroleum asphalt or coal tar pitch, coarse aggregate, fine aggregate, and mineral filler or portland cement (77).

Physical Specifications. - Coarse stone aggregate for membrane water-proofing in Illinois (77) should pass five cycles of the sodium sulfate soundness test (ASTM designation, C 88) with a weighted loss of not more than 15 percent and should be well graded. All of the aggregate should pass a three-eighth-inch sieve and 0 to 1 percent a No. 200 sieve. The weighted loss of fine stone aggregate for the same use should not be more than 10 percent after five cycles of the sodium sulfate soundness test. It should all pass a No. 8 sieve and 0 to 3 percent a No. 200 sieve.

MINERAL FEEDS FOR LIVESTOCK

Use. - Pulverized limestone is used as a source of calcium in mineral feeds for livestock.

Chemical Specifications. - A high-calcium limestone containing more than 95 percent calcium carbonate is commonly recommended (115). The fluorine content of the stone should be low, as rations containing approximately 0.03 percent fluorine from rock phosphate or sodium fluoride were harmful to swine (91) and smaller amounts of fluorine also have been found detrimental to dairy animals (141).

Physical Specifications. - Stone ground to pass 200-mesh or finer is used.

MONOCALCIUM PHOSPHATE (124, 126)

Use. - Monocalcium phosphate, a constituent of certain types of baking powders, is made by treating limestone or lime with phosphoric acid.

General Chemical Specifications. - Selected hydrated lime or limestone is used. The limestone should be pure, high-calcium stone.

NATURAL CEMENT

Use. - Limestone or dolomite is calcined below the sintering temperature to make natural cement. Either a rotary or vertical kiln may be employed.

Chemical Specifications. - Limestone or dolomite, containing 13 to 35 percent clayey material (of which silica is 10 to 22 percent) and 4 to 16 percent alumina and iron oxide, is used. The remainder of the stone may be either calcium carbonate or calcium and magnesium carbonate in any proportions (53).

Physical Specifications. - The stone fed to the rotary kiln is pulverized. Lump stone in pieces of about equal size (53) is charged into the vertical kiln; the burned product subsequently is ground to a powder.

Remarks. - Natural cement is intermediate between portland cement and hydraulic lime but differs from the latter in that lumps of it do not slake in water (95). Natural cement was manufactured for sale at three plants in 1958. Most of the production was used for making masonry cement (85).

OIL WELL DRILLING

The sale of limestone for oil well drilling has been mentioned (89).

PAPER

Use. - In the manufacture of chemical wood pulp for making paper in the tower system of the sulfite process, limestone in concrete towers is reacted with sulfur dioxide gas in the presence of water to form calcium bisulfite, $Ca(HSO_3)_2$. This compound, together with sulfurous acid also formed in the tower, is called "sulfite acid" and is used to digest wood chips so that a separation can be made of the cellulose in the wood from other unwanted constituents.

Chemical Specifications. - The limestone should be high-calcium limestone low in magnesium, although a stone containing not more than 8 to 10 percent magnesium carbonate can be used (38). Ferric oxide, alumina, and acid-insoluble material together should be less than about 2 percent (38). The acid-insoluble material should be light colored and settle rapidly. An analysis of a typical limestone shows 53.68 percent CaO (95.77 percent $CaCO_3$) and 1.78 percent MgO (3.72 percent $MgCO_3$), less than 0.5 percent Fe_2O_3 and Al_2O_3, and less than 1 percent SiO_2 (123). Black specks are to be avoided, hence carbonaceous or bituminous limestones may be unsuitable. Mica flakes, graphite flakes, and pyrite also are said to be deleterious (43).

Physical Specifications (38). - The limestone should be in coarse pieces larger than 3 inches in size. Any impurities present should not occur in large particles. In paper mills, which do not have a means of maintaining a constant temperature of the water going through the tower, a denser limestone should be used in summer than in winter to control the amount of calcium bisulfite in the sulfite acid.

Remarks. - In the milk-of-lime system for preparing sulfite acid, a lime containing at least 40 percent magnesia (MgO) is used. A high-calcium lime is used in the alkaline processes of wood pulp treatment for paper manufacturing. Some of the lime so used is made by burning the calcium carbonate cake recovered during the processes; limestone of high purity also is added to the kiln charge to make up for losses (123).

PHARMACEUTICALS

The sale of limestone for pharmaceuticals has been reported (131). A limestone of high purity would seem to be a requisite.

PORTLAND CEMENT

Use. - Limestone (including oystershell) and clay or shale are the major raw materials used for making portland cement (85). Mixtures of limestone and slag or pure limestone and cement rock also are employed (85). When limestone and clay or shale are used, limestone constitutes roughly 75 to 80 percent by weight of the raw material. The remaining 20 to 25 percent usually is shale or clay. The raw materials are finely ground, blended in carefully proportioned amounts, and burned in a kiln. A clinker is formed which is finely ground with the addition of a small amount of gypsum to yield portland cement.

Chemical Specifications. - Limestone used with clay or shale for cement making should commonly contain more than 75 percent calcium carbonate and less than 3 percent magnesia. Some specifications require less than 5 percent magnesium carbonate (155). Phosphorous pentoxide (P_2O_5) should be less than 0.5 percent and sulfur should be low.

White portland cement requires raw materials that are very low in iron. The iron oxide (Fe_2O_3) content of limestone for this purpose probably should be less than about 0.01 percent. Manganese should be low.

The term "cement rock" is applied to clayey limestone, low in magnesia, that is the major constituent of a portland cement raw mix and needs only small additions of other clayey or calcareous material to adjust the composition of the mix to specifications. Clay or shale are common sources of silica, alumina, and iron oxide and "pure" limestone of calcium carbonate. Other additives may be silica, usually as sand; alumina, as diaspore and bauxite; and iron as pyrite or iron-bearing slag (103).

Physical Specifications. - Chert nodules, other hard masses, or coarse quartz grains are undesirable because they require more than normal grinding to reduce them to a powder (155).

POULTRY GRIT

Use. - Limestone is fed to poultry as a source of calcium for the formation of egg shells and bones. It also serves as a grit or grinding agent in the gizzard (137).

Chemical Specifications. - A high-calcium limestone is desirable. Fluorine content should not exceed 0.1 percent (L. E. Card, personal communication, 1938) because larger amounts may be harmful.

Physical Specifications. - Limestone for poultry grit usually will pass a 4- or a 6-mesh sieve and be retained on a 10-mesh sieve. The grit may be graded by sizes into turkey, chicken, poultry, or bird grit (27).

RAILROAD BALLAST

Use. - Limestone and dolomite are used as ballast for railway tracks.

Specifications. - The American Railway Engineering Association (3) has set specifications for railroad ballast among which are the following

1) Deleterious substances include soft and friable pieces, clay lumps, and material finer than 200-mesh. The first should not exceed 5 percent of the ballast, clay lumps should not exceed 0.5 percent, and minus 200-mesh material should not exceed 1 percent.

2) The percentage of wear, determined with the Los Angeles machine, should not exceed 40 percent except as otherwise specified by the engineer.

3) Ballast for use in regions where freezing temperatures occur must pass the sodium sulfate soundness test with a weighted average loss not exceeding 10 percent after five cycles.

The testing methods used are various procedures specified by the American Society for Testing Materials.

Various size grades of ballast are used (3). The coarsest grade ranges from $2\frac{1}{2}$ inches to 3/4 of an inch, and all of it will pass a 3-inch square-hole sieve, 90 to 100 percent a $2\frac{1}{2}$-inch sieve, 25 to 60 percent a $1\frac{1}{2}$-inch sieve, 0 to 10 percent a 3/4-inch sieve, and 0 to 5 percent a $\frac{1}{2}$-inch sieve. The finest grade specified is a 1-inch to No. 4 sieve grade, all of which will pass a $1\frac{1}{2}$-inch sieve, 95 to 100 percent a 1-inch sieve, 25 to 60 percent a $\frac{1}{2}$-inch sieve, 0 to 10 percent a No. 4 sieve, and 0.5 percent a No. 8 sieve.

RAYONS

The sale of limestones for rayons has been reported (89). This may be a use for limestone whiting.

REFRACTORY DOLOMITE AND DOLOMITE REFRACTORIES

Dolomite is used in three forms as a basic refractory in making steel — as raw dolomite, calcined dolomite, and dead-burned dolomite.

Raw Dolomite

Use. - Raw dolomite is used for patching the hearth of open-hearth furnaces, especially along the slag line (Harley C. Lee, personal communication, 1938), for banking doors, and as a fettling material with dead-burned dolomite (108).

Chemical Specifications. - Dolomite should contain more than 20 percent magnesia, less than 0.05 percent sulfur, and preferably less than 2 percent silica (108). Other specifications (41) call for the dolomite to contain more than 21 percent magnesia, less than 1 percent silica, and less than 0.5 percent alumina.

Physical Specifications. - The dolomite generally is screened to pass a three-fourths-inch or smaller screen and the fines are removed (108). A stone that does not disintegrate when heated is desirable.

Calcined Dolomite

Use. - Calcined dolomite, sometimes called single-burned dolomite (41), is made by heating dolomite to somewhat above 2000°F in suitable kilns. Because calcined dolomite slakes readily in the presence of water, it is usually made at or near the consuming plants (108). The use of calcined dolomite is said to be decreasing (41).

Specifications. – A dolomite meeting the chemical specifications for raw dolomite mentioned above probably would be suitable for making calcined dolomite.

Dead-Burned Dolomite

Use. – Dead-burned dolomite is made by heating to about 3000 °F a mixture of dolomite and iron oxide (108). Other names sometimes applied to dead-burned dolomite are sintered, clinkered, double-burned, and roasted dolomite (108). The amount of iron oxide used varies from 2 to 12 percent and averages about 7 percent (132). Dead-burned dolomite is used in large quantities for patching and mainte-nance work in basic open-hearth furnaces, and a considerable tonnage also is used for electric furnace bottoms (Harley C. Lee, personal communication, 1938). It is likewise used in basic Bessemer converters, crucibles for lead blast furnaces, and in crucibles for melting metals (72).

Chemical Specifications. – Dolomite to be dead burned for use in basic bottoms of open-hearth furnaces should contain less than 1 percent silica, less than 1.5 percent combined alumina and ferric oxide, and at least 35 percent mag-nesium carbonate (21). Other specifications (108) state that dolomite for making dead-burned dolomite should contain "preferably 21 percent or more magnesia, less than 1 percent of silica, and less than 0.5 percent alumina." Another maxi-mum for silica is given as $1\frac{1}{2}$ percent (57).

Physical Specifications. – The dolomite is crushed to relatively small sizes. One specification calls for pieces passing one-half inch (40). An Ohio dolomite plant uses three sizes of washed dolomite: a minus seven-sixteenths-inch sieve product, a minus three-sixteenths-inch sieve product, and a minus five-sixty-fourths-inch sieve product (108).

RICE MILLING

The sale of limestone for rice milling has been reported (89).

RIPRAP

Use. – Riprap consists of large blocks of stone used for foundations and for filling around the base of structures subjected to erosion such as the bases of piers, trestles, abutments, and the like. It also is placed on the banks of streams or the the shores of lakes and other bodies of water to prevent erosion.

Physical Specifications. – No generally accepted specifications for riprap have been noted; however, a weather resistant stone free from cracks, laminae, or other structures which will cause it to split or spall is desirable. Pyrite veinlets, clay partings, or chert generally are undesirable. The Illinois State Division of Highways (82) requires that stone riprap "when subjected to 5 cycles of the sodium sulfate soundness test, shall show no disintegration and not more than 10 percent of the pieces may show splitting, crumbling or spalling." A further requirement is that no stone be less than 6 inches in its smallest dimension (82).

ROCK DUSTING

Use. – Limestone or dolomite dust is applied to the walls, roofs, and floors of coal mines to prevent or check coal dust explosions (70).

Specifications. – The dust should be light colored (16), and to comply with "American Standard practice for rock-dusting underground bituminous-coal and lig-nite mines to prevent coal-dust explosions" (147), it should be

a) a material 100 percent of which will pass through a U.S. No. 20 sieve and 70 percent or more of which will pass through a U. S. No. 200 sieve,

b) a material the particles of which when wetted and dried will not cohere to form a caked dust, and

c) a material that does not contain more than 5 percent of combustible matter or more than a total of 5 percent of free and combined silica (SiO_2). Limestone and similar carbonates produce the best rock dust, as this dust has low silica content, little tendency to cake, and a light color that aids illumination.

ROCK WOOL (104, 107)

Use. - The term "rock wool" was at one time applied largely to mineral wool made from rock, but in recent years it also has been applied to mineral wool made from slag. Only small tonnages of limestone or dolomite currently are sold for making mineral wool. The specifications for limestone and dolomite for making rock wool are given briefly below as a matter of reference.

Chemical Specifications. - Limestones or dolomites for making rock wool, often called "woolrocks," should contain roughly between 20 and 30 percent carbon dioxide, which is equivalent to approximately 45 to 65 percent calcium carbonate, or calcium carbonate and magnesium carbonate. The balance of the woolrock should be principally silica, or silica and alumina. Magnesium carbonate is not necessary in woolrocks although some operators prefer it. The same applies to alumina. Iron sulfide is undesirable.

Physical Specifications. - The size of the rock used depends in a large measure on several variables involved in the operation of the cupola. It generally is desirable that the stone be broken into pieces as small as can be used without restricting the blast through the cupola to a point of inefficiency; from 2 to 5 inches is probably the most common size.

A woolrock in which the noncalcareous constituents occur in small particles well distributed through the rock mass, rather than in large masses, is preferable because such an intimate mixture facilitates melting.

Remarks. - Rock wool can be made from mixtures of limestone or dolomite with shale, sandstone, or other siliceous or aluminous rocks. The mixture should meet the chemical and physical specifications given above.

SILICONES

The sale of limestone for silicones has been reported (88).

STONE CHIPS

Use. - Limestone and dolomite chips are used for stucco, terrazzo aggregate, roofing "gravel," roofing chips, as a facing material for concrete and concrete products, and in the making of artifical stone. Chips also are used in bituminous concrete (p. 5).

Chemical Specifications. - No chemical specifications have been noted.

Physical Specifications. - For terrazzo chips, uniform color and hardness, durability, toughness, ability to take a polish, and low liquid absorption are major requirements (19). Three sizes of terrazzo chips have been mentioned (86): those passing a one-fourth-inch sieve and retained on a one-eighth-inch sieve; those passing a three-eighths-inch sieve and retained on a one-fourth-inch sieve; and those passing a one-half-inch sieve and retained on a three-eighths-inch sieve.

For uses other than terrazzo, the chips should be clean and free from dust, soft particles, shale, and pyrite. Common sizes are thought to range between one-eighth and one-half inch. The chips, especially those to be used for roofing, should have good weather resistance. An attractive color is important for all uses though it may be less important for some types of roofing chips.

STUDIO SNOW

The sale of limestone for studio snow, probably for use in motion picture studios, is reported (47). Pulverized limestone or limestone granules probably are used for this purpose.

SUGAR

Use (14, 146). – Raw juice from sugar beets is agitated with milk of lime, and carbon dioxide is then passed through the juice. Impurities present are carried down by the precipitated calcium carbonate that forms which is then removed by thickening and filtration. The lime and carbon dioxide are made by burning limestone at the sugar factory. In another procedure for processing sugar beet juice, the Steffen process, pulverized lime is used to remove sugar from molasses that results during the sugar recovery process. The insoluble calcium saccharate so formed is recovered, washed, and returned to an earlier stage of the process. The lime so used is burned at the factory. Lime also is used to precipitate impurities from cane sugar juice but is generally not made at the plant (29).

Chemical Specifications. – A limestone of high purity is desired. Minimum calcium carbonate should be 96 percent (71) or 97 percent (145). Silica should not exceed 1 percent and magnesia not more than 1 to 4 percent (39, 63). Another specification for silica sets a maximum of 4 percent (71). Iron oxide is undesirable (63) and at some plants must not exceed 0.5 percent (39). Specifications for limestone used in California sugar factories not using the Steffen process call for 95 percent or more calcium carbonate, 4 percent or less magnesia, and less than 1 percent silica (14).

Physical Specifications. – A considerable variety of sizes are specified including pieces 2 to 6 inches (63); 6 to 8 inches (145); 2 to 4 inches, 4 to 6 inches, or 4 to 8 inches (14); and pieces that pass through 6-inch mesh and are retained on 3-inch mesh (58). Limestone for the making of lime for non-Steffen factories should be fine grained and usually fine grained for Steffen factories (14). Dense, extremely fine-grained limestone is reported to offer difficulties in the Steffen process (71). For non-Steffen factories fossiliferous limestones of marine origin are said to be desirable for lime burning (71). The limestones generally should retain their shape during burning.

SULFURIC ACID PURIFICATION

The sale of limestone for use in the purification of sulfuric acid is reported (143). No specifications have been found covering this use.

TARGET SHEETS

The sale of limestone for target sheets has been noted (88).

WATER TREATMENT

Use. - Limestone is used as a coagulent or stabilizer to prevent afterprecipitation of calcium carbonate from lime-softened water (54, 69, 138).

Chemical Specifications. - Limestones differing somewhat in composition have been used in tests and gave apparently similar results (138). A municipal water works is using crystalline high-calcium limestone (54).

Physical Specifications. - Suggested size specifications vary. Pulverized limestone between 150- and 325-mesh (69), finely ground limestone (54), and limestone ground so that 90 percent will pass a 100-mesh sieve and the remainder a 60-mesh sieve (138) are suggested sizes.

WHITING (LIMESTONE)

Limestone whiting, also referred to as whiting substitute, is made by pulverizing limestone, dolomite, marble, vein calcite, marl, or oyster shells (27, 87). Generally, limestone whiting should be of "good white color" (24) though in the manufacture of some darker colored products whiting of inferior color can be used (154). Although it is sometimes difficult to draw a sharp separation between whiting and other calcium carbonate powders, whiting, to deserve the name, "must be white or nearly so, and should be ground to pass 200-mesh or finer" (24).

The term "whiting" was originally applied to powders made from chalk, but at present it includes limestone whiting as well as precipitated calcium carbonate powders.

Below is a list of uses in which or for which whiting is used. It is compiled from various sources that do not always differentiate the type of whiting referred to. It is probable, however, that limestone whiting will serve for many or most of the purposes mentioned (23, 24, 66, 89, 96, 99, 153).

Acoustic tile
Asbestos products filler (131)
Asphalt (may be off-color whiting)
Calcimine and cold water paints
Caulking compounds
Ceramic glazes, enamels, and bodies
Chemical manufacture
Chewing gum
Cigarette papers
Coating on glazed paper
Cosmetics
Crayons
Dolls
Dressing for white shoes
Dusting unburned brick to prevent
  sticking in the kiln (153)
Dusting printing rollers (129)
Dyes
Explosives
Fabric filler
Facing for molds and cores in brass
  casting
File manufacture

Fireworks
Flavoring extracts (31)
Floor coverings
Foundry compounds
Glue
Graphite filler (35)
Grease
Gypsum plaster
Insecticides
Leather goods
Linoleum
Locomotive works
Magazine and book papers
Making buff brick from red-burning
  clay (66)
Manufacture of citric acid (95)
Medicines
Metal polish
Neutralizing in fermentation processes
Oilcloth
Paints
Paper
Parting compounds (32)

Phonograph records
Picture frame moldings
Plastics
Pottery
Printing ink (49)
Printing and engraving
Putty
Roofing cement
Rubber
Rubber goods (footwear, heels, hard
  rubber objects, white rubber stock,
  molded rubber goods, sponge rubber,
  hose, belts, and mats and electric
  cable insulation [66])

Sealing wax
Ship building
Shoe manufacturing
Soap
Structural iron making
Toothpaste
Welding electrode coatings
White ink
Whitewash (129)
Window shades
Wire insulation

Specifications, either physical or chemical, or both, have been found for some of the uses mentioned above. These are given subsequently by use and are believed to include most, if not all, the major uses and some of the minor uses of limestone whiting. However, for some uses, tests in addition to those mentioned also are made, and in many cases the suitability of a limestone whiting for a particular use is best determined by submitting samples to prospective consumers for those tests which they have found most effective in determining the specific properties of the whiting they require.

Ammonium Nitrate

Limestone or precipitated calcium carbonate is mixed with ammonium nitrate to prevent caking of the latter compound. This practice is prevalent in Europe (136). Presumably a pulverized high-calcium limestone would be used for this purpose.

Ceramic Whiting

Use. - Levigated limestone whiting is used to some extent in making pottery and porcelain ware (25). Limestone whiting also is a constituent of glazes and enamels (24). Pulverized dolomite marble is used (120) as "a body constituent by the vitreous dinnerware producers" and in glaze batches. Experiments indicate that pulverized dolomite marble has value as an auxiliary flux in suitable floor tile compositions (120). It has been found also that in laboratory experiments additions of substantial amounts of finely pulverized limestone or of larger amounts (5 and 10 percent) of pulverized dolomite had a stabilizing effect on the fired properties of a No. 5 fire clay and a shale (55).

Specifications. - The American Ceramic Society (2) specifies two classes of ceramic whiting. Both classes should contain not less than 97 percent total carbonates, nor more than 0.25 Fe_2O_3, 2.0 percent SiO_2, and 0.1 percent SO_3. Class 1 should contain 96 percent or more $CaCO_3$ and not over 1 percent $MgCO_3$. Class 2 may have as much as 8 percent $MgCO_3$ and as little as 89 percent $CaCO_3$.

The whitings used in the laboratory experiments mentioned above (55) were limestone, 80 percent through 200-mesh, $CaCO_3$, 89 percent, $MgCO_3$, 8 percent; and dolomite, pulverized; CaO, 22 percent, MgO, 30 percent.

Filler in Plastics

Calcium carbonate is used as a filler in vinyl compounds and in polyester resins (144).

Floor Covering Filler (87)

Use. - Ground limestone flour is used as a filler in floor coverings such as linoleum.

Specifications. - Most of the ground limestone used for floor coverings is of such size that more than 90 percent will pass a 325-mesh sieve, although coarser grades are used in some products.

Insecticide Diluent

Ground limestone, chemically treated to make it free flowing and nonwettable, has been found to be satisfactory for the formulation of insecticide dusts (125). The limestone should be finely ground. The size specification for many diluents is said to be 95 percent or more through 325-mesh (44 microns) or 95 percent through 200-mesh (74 microns) (119).

Paint Pigments and Filler

Use. - Whiting is used as a pigment and/or filler in paints.

Chemical Specifications. - The whiting may be made from either limestone, dolomite, or magnesian limestone. The limestone should contain not less than 98 percent calcium carbonate (dry basis); dolomite or magnesian limestone should contain not less than 95 percent calcium carbonate and magnesium carbonate (dry basis) (10). Other specifications (39) state that "in general calcium carbonate content should exceed 96% but magnesian limestones containing as much as 8% magnesium oxide are (rarely) tolerated — the $MgCO_3$ content is generally 1%. Other maxima are: Fe_2O_3 - 0.25%, SiO_2 - 2.0% and SO_3 - 0.1%."

Physical Specifications. - Paint-grade limestone whiting should contain not more than 1 percent of material retained on a No. 325 sieve and be largely between 15 to 20 microns in maximum size (10). Filler should not have more than 15 percent retained on a No. 325 sieve with substantial amounts in the 10- to 44-micron particles (10). A main controlling characteristic is the degree of whiteness of the whiting (39). Some plants prefer a stone which breaks into rhombic particles (39). Oil absorption is another important property.

Paper Filler

Use. - Calcium carbonate fillers are widely used in magazine and bible papers (151). Both limestone whiting and by-product precipitated calcium carbonate are used as paper fillers (133). Cigarette papers contain a special precipitated calcium carbonate (151).

Specifications. - No chemical or size specifications for limestone whiting were noted. Stone used for making paper filler probably should be white, have high purity, and be ground to a fine powder. Willets (151) states that limestone whiting for paper is separated by various air-floating and wet techniques.

Putty

The essential ingredients of putty are whiting and linseed oil (154). Limestone whiting is much used for this purpose in the United States. Among other desirable properties are low oil absorption and a light gray or cream color (154). The ASTM specifies that putty-powder grade whiting should not have more than 30 percent retained on a No. 325 sieve (10). Fineness of grain size and freedom from grit are desirable.

Rubber Filler

Use. - Pulverized limestone is used as a filler in rubber of various sorts.

Specifications. - Wilson (154) gives the following ranges in specifications from the poorest to the best grades of whiting for rubber filler:

Color, variable for different products, same as type sample; particle size, 100 percent finer than 100-mesh and 99 to 99.8 percent finer than 325-mesh minimum; ignition loss, 40 to 46 percent minimum; calcium carbonate, 90 to 95 percent minimum; iron, 1.0 to 0.3 percent maximum; manganese, 0.04 to 0.004 percent maximum; moisture, 0.4 to 0.1 percent maximum; and alkalinity, 1.0 to 0.01 percent maximum.

The specifications given may cover precipitated calcium carbonate as well as pulverized limestone.

Welding Electrode Coatings

Powdered marble commonly is used in the coatings of welding electrodes, usually in amounts greater than 5 percent of the weight of the wet batch (51). Typical particle sizes for pulverized marble are (51):

Finer than	Coarser than	Coarse grade	Fine grade
Mesh		%	%
	48	2	0
48	100	45	0
100	200	45	10
200	325	5	30
325		3	60

Presumably limestone whiting of suitable character might be similarly used.

YARDS, PLAYGROUNDS, TENNIS COURTS, ETC.

Use. - Limestone or dolomite is used for tennis courts, yards, playgrounds, station platforms, and the like.

General Physical Specifications. - Stone for these uses should have good weather resistance. No generally adopted size specifications are known. Screenings are said to give satisfactory service (27). However, for many surfacing purposes, such as those mentioned, "small amounts of $\frac{1}{2}$-inch or even $\frac{3}{4}$-inch stone mixed with the finer sizes gives a better body to the mass of stone without in any way injuring its packing qualities or smooth surface. This feature is best appreciated in wet weather" (W. R. Sanborn, personal communication, 1937).

BIBLIOGRAPHY

1. American Association of State Highway Officials, 1955, Standard specifications for highway materials and methods of sampling and testing, Part I [7th ed.]: Am. Assoc. State Highway Officials Specif., p. 52-74, Washington, D. C.

2. American Ceramics Society, 1928, Standard specifications for materials. Ceramic whiting: Am. Ceramics Soc. Jour., v. 11, no. 6, p. 378.

3. American Railway Engineering Association, 1959, Specifications for prepared stone, slag and gravel ballast, in Manual for Railway Engineering: Am. Ry. Eng. Assn., Chicago, p. 1-2-2 to 1-2-4.

4. American Society of Civil Engineers, 1937, Filtering materials for sewage treatment plants, in Manual of Engineering Practice, no. 13, p. 9: Am. Soc. Civil Eng., New York.

5. American Society for Testing Materials, 1958, Tentative method of test for combined effect of temperature cycles and weak salt solutions on natural building stones. ASTM Designation C 218-48T, in 1958 Book of Standards, Part 5, p. 660: ASTM, Philadelphia, Pa.

6. American Society for Testing Materials, 1958, Standard method of test for abrasion of coarse aggregate by the use of the Los Angeles machine. ASTM Designation C 131-55, in 1958 Book of Standards, Part 4, p. 490-492: ASTM, Philadelphia, Pa.

7. American Society for Testing Materials, 1958, Abrasion resistances of stone subjected to foot traffic. ASTM Designation C 241-51, in 1958 Book of Standards, Part 5, p. 648-650: ASTM, Philadelphia, Pa.

8. American Society for Testing Materials, 1958, Cement, concrete, mortars, road materials, waterproofing, soils, in 1958 Book of Standards, Part 4, p. 446-462, 483-489: ASTM, Philadelphia, Pa.

9. American Society for Testing Materials, 1958, Standard definitions of terms relating to concrete and concrete aggregates. ASTM Designation C 125-58, in 1958 Book of Standards, Part 4, p. 588-589: ASTM, Philadelphia, Pa.

10. American Society for Testing Materials, 1958, Tentative specifications for calcium carbonate pigments. ASTM Designation D 1199-52T, in 1958 Book of Standards, Part 8, p. 11-12: ASTM, Philadelphia, Pa.

11. American Society for Testing Materials, 1958, Tentative specifications for mineral filler for sheet asphalt and bituminous concrete pavement. ASTM Designation D 242-57T, in 1958 Book of Standards, Part 4, p. 450: ASTM, Philadelphia, Pa.

12. American Society for Testing Materials, 1959, Tentative method of test for soundness of aggregates by use of sodium sulfate or magnesium. ASTM Designation C 88-59T, in 1958 Book of Standards Supplement, Part 4, p. 91: ASTM, Philadelphia, Pa.

13. Arundale, J. C., 1956, Sodium compounds, in Mineral Facts and Problems: U. S. Bur. Mines Bull. 556, p. 797.

14. Ballou, F. H., Jr., 1951, Limestone in the California beet sugar industry: California Jour. Mines and Geology, California Div. Mines, v. 47, no. 1, p. 10-16.

15. Beeson, K. C., and Ross, W. H., 1935, Preparation of physiologically neutral fertilizer mixtures: Ind. Eng. Chem., Industrial edition, v. 26, no. 9, p. 992.

16. Bierer, Joseph, 1948, Rock dusting in coal mines — A must in protecting lives and property: Pit and Quarry, v. 42, no. 3, p. 92.

17. Birch, R. E., and Wicken, O. M., 1949, Magnesite and related minerals, *in* Industrial Minerals and Rocks. p. 529: Am. Inst. Min. Met. Eng., New York.

18. Blue, D. D., 1954, Raw materials for aluminum production: U. S. Bur. Mines Inf. Circ. 7675, p. 2, 5.

19. Bowen, Oliver, Jr., 1957, Recent developments in limestones, dolomites and cements in California: Mining Cong. Jour., v. 43, no. 8, p. 79.

20. Bowles, Oliver, 1928, Utilization problems of metallurgical limestone and dolomite: Am. Inst. Min. Met. Eng. Tech. Pub. 62, p. 9, 12, 13.

21. Bowles, Oliver, 1928, Utilization problems of metallurgical limestone and dolomite: Am. Inst. Min. Met. Eng. Tech. Pub. 62, p. 14.

22. Bowles, Oliver, 1929, Metallurgical limestones, problems in production and utilization: U. S. Bur. Mines Bull. 299, p. 19-22.

23. Bowles, Oliver, 1931, Chalk, whiting and whiting substitutes: U. S. Bur. Mines Inf. Circ. 6482, p. 7-8.

24. Bowles, Oliver, 1942, Chalk and whiting: U. S. Bur. Mines Inf. Circ. 7197, p. 3-6.

25. Bowles, Oliver, 1941, The occurrence and uses of dolomite in the U. S.: U. S. Bur. Mines Inf. Circ. 7192, p. 9.

26. Bowles, Oliver, 1952, The lime industry: U. S. Bur. Mines Inf. Circ. 7651, p. 2, 36.

27. Bowles, Oliver, 1956, Limestone and dolomite: U. S. Bur. Mines Inf. Circ. 7738, p. 13-14.

28. Bowles, Oliver, and Banks, D. M., 1933, Limestone. Part I — General information: U. S. Bur. Mines Inf. Circ. 6723, p. 12.

29. Bowles, Oliver, and Banks, D. M., 1936, Lime: U. S. Bur. Mines Inf. Circ. 6884, p. 32.

30. Bowles, Oliver, and Jensen, Mabel S., 1941, Limestone and dolomite in the chemical and processing industries: U. S. Bur. Mines Inf. Circ. 7169, p. 6.

31. Bowles, Oliver, and Jensen, Mabel S., 1943, Stone, *in* Minerals Yearbook, 1942, p. 1249: U. S. Bur. Mines.

32. Bowles, Oliver, and Jensen, Mabel S., 1943, Stone, *in* Minerals Yearbook, 1941, p. 1252: U. S. Bur. Mines.

33. Bowles, Oliver, and Jensen, Mabel S., 1945, Stone, *in* Minerals Yearbook, 1943, p. 1307: U. S. Bur. Mines.

34. Bowles, Oliver, and Jensen, Nan C., 1947, Industrial uses of limestone
    and dolomite: U. S. Bur. Mines Inf. Circ. 7402, p. 11, 15.

35. Bowles, Oliver, and Schauble, M., 1939, Stone, *in* Minerals Yearbook,
    1939, p. 1144: U. S. Bur. Mines.

36. Boynton, Robert S., and Gutschick, Kenneth A., 1960, Lime, *in* Industrial
    Minerals and Rocks, p. 502: Am. Inst. Min. Met. Petroleum Eng., New
    York.

37. British Standards Institution, 1959, Limestone for making colourless glasses:
    British Standards Inst. Specification BS 3108, p. 5.

38. Cadigan, A. M., 1942, Limestone and lime in the chemical pulp industry:
    Pit and Quarry, v. 35, no. 5, p. 74-78.

39. California Division of Mines, 1959, Limestone, dolomite and lime products:
    California Div. Mines Mineral Inf. Serv., v. 12, no. 2, p. 8.

40. Camp, J. M., and Francis, C. B., 1925, The making and shaping of steel
    [4th ed.], p. 295: Carnegie Steel Company, Pittsburgh, Pa.

41. Camp, J. M., and Francis, C. B., 1957, The making, shaping and treating
    of steel [7th ed.], p. 172, 173, 181, 184, 185: Carnegie Steel Company,
    Pittsburgh, Pa.

42. Chemical Engineering, 1958, Magnesia from sea via streamlined process:
    Chem. Eng., v. 65, no. 6, p. 112.

43. Claudet, H. H., 1928, Limestone for the pulp and paper industry: Canadian
    Mining Jour., v. 49, April, p. 306.

44. Colby, Shirley F., 1941, Occurrence and uses of dolomite in the United
    States: U. S. Bur. Mines Inf. Circ. 7192, p. 8.

45. Comstock, H. B., 1956, Magnesium, *in* Mineral Facts and Problems: U. S.
    Bur. Mines Bull. 556, p. 476-484.

46. Comstock, H. B., 1957, Magnesium: Preprint, Bureau of Mines Yearbook,
    1957, U. S. Bur. Mines, p. 2.

47. Coons, A. T., 1935, Stone, *in* Minerals Yearbook, 1934, Statistical Appendix,
    p. 54: U. S. Bur. Mines.

48. Cooper, J. E., 1950, How to dispose of acid wastes: Chem. Ind., v. 66,
    no. 5, p. 685.

49. Cummins, Arthur B., 1960, Mineral fillers, *in* Industrial Minerals and Rocks,
    p. 578, 580: Am. Inst. Min. Met.Petroleum Eng., New York.

50. Currier, L. W., 1960, Geologic appraisal of dimension-stone deposits:
    U. S. Geol. Survey Bull. 1109, p. 26-39.

51. DeLong, W. T., and Reed, H. F., Jr., 1955, Ceramic raw materials for the
    welding electrode industry: Am. Ceramics Soc. Bull., v. 34, no. 9, p.
    183-184.

52. Eardley-Wilmot, V. L., 1927, Abrasives. Part I — Siliceous abrasives:
    Canada Dept. Mines, Mines Branch, no. 673, p. 97.

53. Eckel, E. C., 1928, Cements, limes and plasters: John Wiley & Sons, New York, p. 206, 218.

54. Engineering News-Record, 1957, Small waterworks features six major innovations: Eng. News-Rec., v. 159, no. 17, p. 32, 33.

55. Everhart, J. O., 1957, Use of auxiliary fluxes to improve structural clay products: Am. Ceramics Soc. Bull., v. 36, no. 7, p. 269.

56. Gault, H. R., and Ames, John, 1960, Fluxing stone, in Industrial Minerals and Rocks [3rd ed.], p. 191: Am. Inst. Min. Met. Petroleum Eng., New York.

57. Gibbs, Ralph, 1949, Manufacturing refractory dolomite: Rock Products, v. 52, no. 4, p. 129.

58. Giddings, Ray C., 1954, Producing high grade "sugar" stone: Rock Products, v. 57, p. 60.

59. Gillson, J. L., 1960, The carbonate rocks, in Industrial Minerals and Rocks, p. 194: Am. Inst. Min. Met. Petroleum Eng., New York.

60. Goldbeck, A. T., 1949, Crushed rock, in Industrial Minerals and Rocks, p. 262, 263,264: Am. Inst. Min. Met. Petroleum Eng., New York.

61. Goldbeck, A. T., 1949, as reported in Crushed stone production 1949 to equal 1948 record volume: Rock Products, v. 52, no. 3, p. 110.

62. Goudge, M. F., 1930, Limestone in industry, in Investigations of Mineral Resources and the Mining Industry, 1929: Canada Dept. Mines, Mines Branch, no. 719, p. 46.

63. Goudge, M. F., 1930, Limestone in industry: Reprint from the Canadian Min. and Met. Bull., from Canadian Inst. Min. Met. Trans., 1930, p. 4, 6, 7.

64. Goudge, M. F., 1930, Limestone in industry, in Investigations of Mineral Resources and the Mining Industry, 1929: Canada Dept. Mines, Mines Branch, no. 719, p. 45, 46.

65. Goudge, M. F., 1937, Limestone and lime — Their industrial uses: Mining and Metallurgy, v. 18, no. 368, p. 372.

66. Goudge, M. F., 1937, Limestone and lime — Their industrial uses: Mining and Metallurgy, v. 18, no. 368, p. 373-374.

67. Grindrod, John, 1959, Dolomite + sea water = refractory magnesia: Pit and Quarry, v. 52, no. 3, p. 102-106.

68. Harness, Charles L., and Jensen, Nan C., 1943, Magnesium compounds and miscellaneous salines, in Minerals Yearbook, 1942, p. 1498: U. S. Bur. Mines.

69. Harting, Herbert O., 1956, Calcium carbonate stabilization of lime-softened water: Am. Waterworks Assoc. Jour., v. 48, no. 12, p. 1523, 1534.

70. Hartman, Irving, and Westfield, James, 1956, Rock dusting and sampling: U. S. Bur. Mines Inf. Circ. 7755, p. 2.

71. Hartmann, E. M., 1951, Lime and carbon dioxide production, in Beet-sugar technology (R. A McGinnis, ed.), p. 427-428: Reinhold Publishing Corporation, New York.

72. Hatmaker, Paul, 1931, Utilization of dolomite and high-magnesium limestone: U. S. Bur. Mines Inf. Circ. 6524, p. 3.

73. Heath, W. P., 1930, Manufacture of carbon dioxide and epsom salts: Ind. Eng. Chem., v. 22, p. 437.

74. Hoak, Richard D., Lewis, C. J., and Hodge, W. W., 1945, Treatment of spent pickling liquors with limestone and lime: Ind. Eng. Chem., v. 37, no. 6, p. 556.

75. Hoak, Richard D., Lewis, Clifford J., Sindlinger, Charles J., and Klein, Bernice, 1948, Pickle liquor neutralization: Ind. Eng. Chem., v. 40, no. 11, p. 2065.

76. Huntley, Victoria H., 1960, On making a lithograph: Am. Artist, v. 24, no. 5, p. 30-35.

77. Illinois Division of Highways, 1958, Standard specifications for road and bridge construction, p. 360, 363, 364, 608, 621: Illinois Div. Highways, Springfield.

78. Illinois Division of Highways, 1958, Standard specifications for road and bridge construction, p. 105, 111, 125, 143, 149, 154, 161, 180, 192, 203, 213, 220, 229, 250, 268, 604-622: Illinois Div. Highways, Springfield.

79. Illinois Division of Highways, 1958, Standard specifications for road and bridge construction, p. 710: Illinois Div. Highways, Springfield.

80. Illinois Division of Highways, 1958, Standard specifications for road and bridge construction, Division III — Material Details, p. 604: Illinois Div. Highways, Springfield.

81. Illinois Division of Highways, 1958, Standard specifications for road and bridge construction, Division III — Material Details, p. 611: Illinois Div. Highways, Springfield.

82. Illinois Division of Highways, 1958, Standard specifications for road and bridge construction, Division III — Material Details, p. 622, 623: Illinois Div. Highways, Springfield.

83. Jacob, K. D. [ed.], 1953, Fertilizer technology and resources in the U. S., v. 3, p. 325: Academic Press, Inc., New York.

84. Jones, Edward M., 1950, Acid wastes treatment: Sewage and Industrial Wastes, v. 22, no. 2, p. 224-227.

85. Kennedy, D. O., and Moore, Betty M., 1959, Cement, in Minerals Yearbook, 1958, p. 281, 287: U. S. Bur. Mines.

86. Kessler, D. W., Hockman, A., and Anderson, R. E., 1943, Physical properties of terrazzo aggregates: U. S. Natl. Bur. Standards BMS 98, p. 5, 6.

87. Key, Wallace W., 1960, Chalk and whiting, in Industrial Minerals and Rocks [3rd ed.], p. 233, 240: Am. Inst. Min. Met. Petroleum Eng., New York.

88. Key, Wallace W., and Jensen, Nan C., 1956, Stone, in Minerals Yearbook, 1956, v. 1, p. 1109: U. S. Bur. Mines.

89. Key, W. W., and Jensen, N. C., 1959, Stone, *in* Minerals Yearbook, 1958, v. 1, p. 990: U. S. Bur. Mines.

90. Key, W. W., Holmes, George H., Jr., and Jensen, N.C., 1960, Stone, *in* Minerals Yearbook, 1959, v. 1, p. 1013: U.S. Bur. Mines.

91. Kirk, C. H., and Bethke, R. M., 1933, Effect of fluorine on the nutrition of swine with special reference to bone and tooth composition: Agr. Res. Jour., v. 46, no. 11, p. 1035.

92. Kirk, R. E., and Othmer, D. F. [eds.], 1947, Encyclopedia of chemical technology: Interscience Encyclopedia, Inc., New York, v. 1, p. 61.

93. Kirk, R. E., and Othmer, D. F. [eds.], 1948, Encyclopedia of chemical technology: Interscience Encyclopedia, Inc., New York, v. 2, p. 766.

94. Kirk, R. E., and Othmer, D. F. [eds.], 1949, Encyclopedia of chemical technology: Interscience Encyclopedia, Inc., New York, v. 3, p. 127.

95. Knibbs, N. V. S., 1924, Lime and magnesia, p. 240: D. Van Nostrand Company, New York.

96. Knibbs, N. V. S., 1924, Lime and magnesia, p. 269: D. Van Nostrand Company, New York.

97. Knibbs, N. V. S., 1924, Lime and magnesia, p. 259: D. Van Nostrand Company, New York.

98. Kriege, Herbert F., 1948, Mineral aggregates in the chemical and processing industries and in certain other industries, *in* ASTM Symposium on Mineral Aggregates: ASTM Spec. Tech. Pub. 83, p. 208.

99. Ladoo, R. B., 1925, Nonmetallic minerals, p. 130: McGraw-Hill Book Company, Inc., New York.

100. Ladoo, R. B., and Myers, W. M., 1951, Nonmetallic minerals [2nd ed.], p. 296: McGraw-Hill Book Company, Inc., New York.

101. Lamar, J. E., 1930, Chert gravel as sewage filter stone: Sewage Works Jour., v. 2, no. 4, p. 495-499.

102. Lamar, J. E., 1950, Acid etching in the study of limestones and dolomites: Illinois Geol. Survey Circ. 156.

103. Lamar, J. E., 1959, Limestone resources of extreme southern Illinois: Illinois Geol. Survey Rept. Inv. 211, p. 16.

104. Lamar, J. E., and Machin, J. S., 1949, Heat and sound insulators, *in* Industrial Minerals and Rocks, p. 448-467: Am. Inst. Min. Met. Eng., New York.

105. Lamar, J. E., and Shrode, R. S., 1953, Water soluble salts in limestones and dolomites: Econ. Geology, v. 48, p. 110.

106. Lamar, J. E., and Willman, H. B., 1938, A summary of the uses of limestone and dolomite: Illinois Geol. Survey Rept. Inv. 49, p. 27, 28.

107. Lamar, J. E., and Willman, H. B., Fryling, C. F., and Voskuil, W. H., 1934, Rock wool from Illinois mineral resources: Illinois Geol. Survey Bull. 61, p. 15, 16, 180, 185, 198.

108. Lee, Harley C., 1947, Refractories from Ohio dolomite: Ohio State Univ. Eng. Exp. Station News, v. 19, no. 2, p. 38-42.

109. Leonard, J. B., 1958, CO_2, a steadily growing giant: Chem. Eng. News, v. 36, no. 40, p. 115.

110. Linsley, C. M., 1954, Limestone, how to use it, when to use it, where to use it: Univ. of Illinois Agr. Ext. Circ. 721, p. 6.

111. Lumsden, J., 1939, Magnesium, magnesite and dolomite, in Reports on the Minerals Industry of the British Empire and Foreign Countries, p. 42: Imperial Institute, London.

112. MacIntire, W. H., Hardin, L. J., and Oldham, F. D., 1936, Phosphate fertilizer mixtures: Ind. Eng. Chem., Industrial ed., v. 28, no. 6, p. 711.

113. Mantell, C. L., 1931, Industrial electrochemistry, p. 405-407: McGraw-Hill Book Company, Inc., New York.

114. Martino, Paul D., and Stone, Robert L., 1956, Limestone-marl mixtures for extruded and dry pressed building brick: Am. Ceramics Soc. Bull., v. 35, July, p. 286-288.

115. McCampbell, C. W., 1931, Feeding ground limestone to cattle: Rock Products, v. 34, no. 14, p. 63.

116. Miller, B. L., 1925, Limestones of Pennsylvania: Pennsylvania Topog. and Geol. Survey Bull. M 7, 4th ser., p. 40.

117. Miller, B. L., 1934, Limestones of Pennsylvania: Pennsylvania Topog. and Geol. Survey Bull. M 20, 4th ser., p. 91.

118. Miller, B. L., 1934, Limestones of Pennsylvania: Pennsylvania Topog. and Geol. Survey Bull. M 20, 4th ser., p. 99, 100.

119. Miller, Roy E., 1950, Agricultural diluents: Agr. Chemicals, v. 5, no. 11, p. 45.

120. Morse, G. T., 1948, Use of dolomite as an auxiliary flux in floor tile: Am. Ceramics Soc. Jour., v. 31, no. 3, p. 67-70.

121. Nicholas, I. A., 1946, in discussion of paper by D. E. Washburn, Limestone and lime for open-hearth use: 29th Open-hearth Conference Proc., v. 29, p. 46, AIME.

122. North, O. S., 1956, Cement, in Mineral Facts and Problems: U. S. Bur. Mines Bull. 556, p. 160.

123. Parsons, John L., 1957, The sulfite process, in Modern Pulp and Paper Making [John B. Calkins, ed.], p. 85, 135, 137: Reinhold Publishing Corporation, New York.

124. Partington, J. R., 1950, Textbook of inorganic chemistry [6th ed.], p. 759: MacMillan and Company, Ltd., London.

125. Rainwater, C. F., Dunnam, E. W., Ivy, E. E., and Scales, A. L., 1953, Calcium carbonate as a diluent for insecticide dusts: Econ. Entomology Jour., v. 46, December, p. 923-927.

126. Riegel, E. R., 1933, Industrial chemistry, p. 130: The Chemical Catalog Company, New York.

127. Riegel, E. R., 1933, Industrial chemistry, p. 435: The Chemical Catalog Company, New York.

128. Riegel, E. R., 1941, Industrial chemistry, p. 221: Reinhold Publishing Corporation, New York.

129. Robertson, R. H. S., 1960, Mineral use guide, p. 12-16: Cleaver-Hume Press Ltd., London.

130. Rogers, Allen, 1931, Industrial chemistry [5th ed.], v. 1, p. 341: D. Van Nostrand Company, New York.

131. Runner, D. G., Jensen, Nan C., and Downey, M. G., 1950, Stone, *in* Minerals Yearbook, 1948, p. 1174: U. S. Bur. Mines.

132. Schallis, Alvin, 1942, Dolomite base refractories: U. S. Bur. Mines Inf. Circ. 7227, p. 4.

133. Schwalbe, H. C., 1955, New developments in the loading of paper: Paper Trade Jour., v. 139, no. 11, p. 33.

134. Searle, Alfred B., 1935, Limestone and its products, p. 527: Ernest Benn, Ltd., London.

135. Searle, Alfred B., 1935, Limestone and its products, p. 251, 604: Ernest Benn, Ltd., London.

136. Shearon, W. H., Jr., and Dunwoody, W. B., 1953, Ammonium nitrate: Ind. Eng. Chem., v. 45, no. 3, p. 496, 497, 502.

137. Smith, R. E., and MacIntyre, T. M., 1959, The influence of soluble and insoluble grit upon the digestibility of feed by domestic fowl: Canadian Jour. Animal Sci., v. 39, December, p. 164-169.

138. Spaulding, Charles H., Lowe, Harry N., Jr., and Schmitt, Richard P., 1951, Improved coagulation by the use of pulverized limestone: Am. Waterworks Assoc. Jour., v. 43, no. 10, p. 796-799.

139. Stewart, A., 1958, Direct reduction of iron ores in a rotary kiln: Min. Cong. Jour., v. 44, no. 12, p. 34-39.

140. Taylor, C. A., and Rinkenbach, W. H., 1923, Explosives, their materials, constitution and analysis: U. S. Bur. Mines Bull. 219, p. 52.

141. Taylor, E. G., 1929, Effect of fluorine in cattle rations: Michigan Agr. Exp. Station Quart. Bull. 11, p. 104.

142. Thiel, George A., and Dutton, Carl E., 1935, Architectural, structural and monumental stones of Minnesota: Univ. of Minnesota, Minnesota Geol. Survey Bull. 25, p. 39.

143. Thoenen, J. R., 1934, Crushed and broken stone, *in* Minerals Yearbook, 1934, p. 831: U. S. Bur. Mines.

144. Thomson, Robert D., 1960, Nonmetallic mineral filters in plastics: Am. Inst. Min. Met. Petroleum Eng. Preprint 60H105, p. 9.

145. Turner, A. M., 1932, Use of limestone in the beet sugar industry: Rock Products, v. 35, no. 19, p. 18.

146. United States Beet Sugar Association, 1959, The beet sugar story, p. 48-51: U. S. Beet Sugar Assoc., Washington, D. C.

147. United States Bureau of Mines, 1960, American standard practice for rock-dusting underground bituminous-coal and lignite mines to prevent coal-dust explosions (ASA standard M13.1-1960, UDC 622.81): U. S. Bur. Mines Inf. Circ. 8001, p. 2.

148. United States Department of Agriculture, 1959, Illinois ACP Specification Sheet: U. S. Dept. Agr., Agr. Stab. and Cons. Prog. Spec. Sheet No. 12.

149. United States General Services Administration, 1953, Interior marble, soapstone, slate, etc.: U. S. Gen. Serv. Admin., Pub. Bldg. Serv., November, p. 2.

150. Whitaker, L. R., 1956, Manufacture of brick and tile from extruded limestone: Am. Ceramics Soc. Bull., v. 35, July, p. 286.

151. Willets, William R., 1952, Paper fillers: Paper Trade Jour., v. 135, no. 20, p. 309.

152. Willman, H. B., 1943, High-purity dolomite in Illinois: Illinois Geol. Survey Rept. Inv. 90, p. 13.

153. Wilson, Hewitt, 1927, Ceramics, clay technology, p. 196: McGraw-Hill Book Company, Inc., New York.

154. Wilson, Hewitt, 1949, Chalk and whiting, in Industrial Minerals and Rocks [2nd ed.], p. 189, 190: Am. Inst. Min. Met. Petroleum Eng., New York.

155. Wolfe, J. A., 1955, What to look for in selecting cement raw materials: Rock Products, v. 58, no. 8, p. 180.

156. Wregs, E. E., and Anstrand, C. J., 1945, Production of magnesium chloride from dolomite: Am. Inst. Chem. Eng. Trans., v. 41, February 25, p. 1-18.

INDEX

ADDENDA

Since the initial publication of this Circular in 1961, a few additional uses or sources of specifications have been noted. These are given below, together with references to the sources from which the information came.

AGRICULTURAL LIMESTONE

Agricultural limestone is said (166) to act as a control to the quantity of strontium 90 removed by plants from soil contaminated by atomic fallout. (See also page 8.)

BURNISHING PEBBLES

Rounded limestone pebbles are used in mills in which castings are burnished. No specifications for the limestone are known. Some pebbles are about one-half or three-fourths inches in diameter. The amount and nature of the impurities in the limestone may well affect their suitability. Presumably dolomite pebbles might be similarly used.

DEAD-BURNED DOLOMITE

Dolomite to be dead-burned (167) usually is of high purity, containing less than 1 percent silica and 0.5 percent alumina. It is used in two sizes, minus three-eighths inch + 6 mesh and -6 mesh + 20 mesh. (See also page 23.)

FARM "TROMP" LOTS

Limestone has been used experimentally (164) as a surfacing material to eliminate boggy tracts caused by the tramping of dairy cows in feeding and watering areas. The stone used was aggregate ranging from 2 inches to dust.

FLUX AND REFRACTORY DOLOMITE

Usual physical specifications for flux and refractory dolomite produced at a recently built steel company stone plant are: blast furnace flux, 4 x 1 inch; open hearth flux, 8 x $2\frac{1}{2}$ inches; rice dolomite and rice calcite, one-half inch x 100 mesh; and sinter flux sand, minus one-fourth inch (162). (See also pages 15, 16, 22.)

MAGNESIUM BISULFITE, MAGNESIUM SULFITE HEXAHYDRATE, AND MAGNESIUM OXIDE

The above compounds result from the leaching of dolomite and brucite with carbon dioxide and sulfur dioxide in a new process (157). The magnesium bisulfite produced is suitable for use in the bisulfite paper-pulping process (157). Presumably high-purity dolomite would be used. (See also pages 18, 19.)

MAGNESIUM COMPOUNDS AND MAGNESIUM

One user of dolomite for making magnesium hydroxide slurry from brines for use by the refractory and pulp and paper industries employs "very high-purity dolomite that contains less than 0.30 percent acid insoluble (mostly silica) and less than 0.30 percent R_2O_3 (Al_2O_3 + Fe_2O_3)". (R. B. Rowe, personal communication, 1964).

A magnesium producing plant in Alabama uses as its ore a dolomitic limestone averaging 18 to 20 percent magnesium carbonate in 1- by one-half-inch pieces (168 and 169). (See also pages 18, 19.)

PAPER

According to Mills (170), the requirements of most paper companies in the state of Washington would be met by a limestone containing not less than 95 percent $CaCO_3$, less than 3.0 percent MgO, less than 1.5 percent silica, and less than 0.5 percent R_2O_3.

Specifications for high calcium limestone and for dolomite to be used in the Jensen Tower System of sulfite pulp production are as follows (163):

High calcium limestone — CaO more than 53 percent; MgO less than 1.5 percent; SiO_2, Fe_2O_3, and Al_2O_3 together less than 1.5 percent; and organic matter less than 0.5 percent.

Dolomite — $CaCO_3$ 54 to 59 percent; $MgCO_3$ 35 to 44 percent; iron and alumina not more than 1 percent, and total insoluble not over 2 percent.

Physical specifications require "man-sized blocks of 8-14 inch diameter."

(See also page 20.)

PIGMENT

Pulverized limestone is used as a pigment in paint, plastics, putties, and wood fillers (161). This is presumably a use for limestone whiting. (See also pages 26-29.)

PREVENTION OF REFLECTION CRACKS

Stone dust passing a 4-mesh sieve has been utilized in a process to prevent reflection cracks in asphalt used to resurface old concrete pavements (158). The kind of stone is not specified.

REFRACTORY DOLOMITE

The A.S.T.M. (160) classification for raw refractory dolomite on an "as received" basis requires a minimum of 16 percent magnesium oxide and gives the following maxima for impurities: silica 1.75 percent; aluminum oxide plus titanium oxide 1.50 percent; and sulfur 0.08 percent. (See also pages 22-23.)

RESIDUAL OIL ADDITIVE

Dolomite has been found to be useful as an additive to residual oil (159).

ROCK DUSTING

According to recent Polish research, the material in rock dust finer than 10 microns is not effective in arresting explosion propagation, and rock dust containing more than 10 percent minus 10 micron material is ineffective (165). (See also page 23.)

SWEEPING COMPOUNDS

The sale of ground limestone to janitor supply houses for use in sweeping compounds has been reported (K. K. Landes, personal communication, 1961).

BIBLIOGRAPHY FOR ADDENDA

157. Anon., 1962, Bringing out bisulfite: Chem. Week, v. 90, no. 26, p. 40.

158. Anon., 1963, Stone dust prevents reflection cracks: Eng. News Rec., v. 170, no. 21, p. 95.

159. Anon., 1962, Technology news letter: Chem. Week, v. 91, no. 14, p. 57.

160. American Society for Testing Materials, 1961, Book of ASTM Standards, Part 5, p. 333, ASTM, Philadelphia, Pa.

161. Ammons, V. G., 1963, The dispersed pigment problem: Industrial and Eng. Chemistry, v. 55, no. 4, p. 42, 46.

162. Bergstrom, J. H., 1963, Bethlehem launches sleek stone plant: Rock Products, v. 66, no. 1, p. 90.

163. Hewitt, D. F., 1960, The limestone industries of Ontario: Ontario Dept. Mines, Industrial Mineral Circ. 5, p. 16.

164. Kennedy, A. B., 1963, Limestone finds new uses: Rock Products, v. 66, no. 10, p. 86, 118, 119.

165. Kingery, D. S., and Mitchell, D. W., 1964, Observation on control of the coal dust explosion hazard in European coal mines: A.I.M.E., Soc. Mining Engr. Trans., v. 229, no. 2, p. 157, 158.

166. Koch, R. M., and Sonneborn, C. B., 1962, The limestone industry's interest in atomic fallout: Pit and Quarry, v. 54, no. 11, p. 168-172.

167. Lee, H. C., 1962, Dead-burned dolomite, its manufacture and use in steel refining furnaces: Am. Ceramics Soc. Bull., v. 41, no. 12, p. 807.

168. Meschter, Elwood, 1961, Aggregate plant readies magnesium ore: Rock
        Products, v. 64, no. 4, p. 89.

169. Meschter, Elwood, 1961, Almet distills limestone to yield magnesium: Rock
        Products, v. 64, no. 4, p. 85-88.

170. Mills, J. W., 1962, High-calcium limestones of eastern Washington:
        Washington Div. Mines and Geology Bull. 48, p. 31.

Illinois State Geological Survey Circular 321
First printing, 1961, 41 p., 2 tables.
Second printing, 1965, with addenda,
44 p., 2 tables

CIRCULAR 321

ILLINOIS STATE GEOLOGICAL SURVEY

URBANA

www.ingramcontent.com/pod-product-compliance
Lightning Source LLC
Chambersburg PA
CBHW080615180526
45168CB00007B/2921